Fowls: Care and Feeding
The Basics of How To Feed and Keep Chickens

by US Dept of Agriculture

with an introduction by Jackson Chambers

This work contains material that was originally published in 1896.

This publication is within the Public Domain and was originally published with Public Funding for the Public Benefit.

This edition is reprinted for educational purposes and in accordance with all applicable Federal Laws.

Introduction Copyright 2017 by Jackson Chambers

Self Reliance Books

Get more historic titles on animal and stock breeding, gardening and old fashioned skills by visiting us at:

http://selfreliancebooks.blogspot.com/

Introduction

I am pleased to present yet another title on Poultry.

The work is in the Public Domain and is re-printed here in accordance with Federal Laws.

As with all reprinted books of this age that are intended to perfectly reproduce the original edition, considerable pains and effort had to be undertaken to correct fading and sometimes outright damage to existing proofs of this title. At times, this task is quite monumental, requiring an almost total "rebuilding" of some pages from digital proofs of multiple copies. Despite this, imperfections still sometimes exist in the final proof and may detract from the visual appearance of the text.

I hope you enjoy reading this book as much as I enjoyed making it available to readers again.

Jackson Chambers

CONTENTS.

	Page.
Introduction	3
Selection of site for buildings and yards	3
Construction of houses	5
Ventilation	7
Perches	8
Nests	9
Drinking fountains	10
Dust boxes	11
Yards or parks	11
Selection of breeds and breeding	12
Feeding	13
Green food	15
Grit	16
Meat food	16
Feeding small chickens	17
Brooders	18
Incubators	19
Disease and lice	21
Dressing and shipping	22

ILLUSTRATIONS.

Fig. 1. Method of building a poultry house with hollow side wall	6
2. Ventilator for a poultry house	8
3. Drinking fountain	10
4. Feeding trough	17

FOWLS: CARE AND FEEDING.

INTRODUCTION.

The wide distribution of domestic fowls throughout the United States and the general use made of their products make poultry of interest to a large number of people. Breeders are continually striving to improve the fowls for some particular purpose, and to excel all predecessors in producing just what the market demands for beauty or utility; but the mass of people look at the poultry products solely as supplying the necessary elements of food in an economical and palatable form. For a considerable time each year eggs are sought instead of meat by people of moderate means, because at the market price eggs are a cheaper food than the various kinds of fresh meat.

Large numbers of the rural population live more or less isolated, and find it inconvenient, if not impossible, to supply fresh meat daily for the table aside from that slaughtered on the farm; and of all live stock poultry furnishes the most convenient means of supplying an excellent quality of food in suitable quantities. This is particularly true during the hot summer months, when fresh meat will keep only a short time with the conveniences usually at the farmer's command.

The general consumption of poultry and poultry products by nearly all classes of people furnishes home markets in almost every city and town in the United States, and at prices which are usually remunerative if good judgment is exercised in the management of the business.

Although fowls require as wholesome food as any class of live stock, they can be fed perhaps more than any other kind of animals on unmerchantable seeds and grains that would otherwise be wholly or partially lost. These seeds often contain various weed seeds, broken and undeveloped kernels, and thus furnish a variety of food which is always advantageous in profitable stock feeding. There is less danger of injury to poultry from these refuse seeds than is the case with any other kind of animals. As a rule, noxious weed seeds can be fed to fowls without fear of disseminating the seeds through the manure, which is not generally true when the weed seeds are fed to other classes of live stock, particularly in any considerable quantity.

SELECTION OF SITE FOR BUILDINGS AND YARDS.

Too often the location of the poultry house is thought to be of minor importance, and consequently is given less consideration than that of any other farm building. Frequently the other buildings are located first and the poultry house then placed on the most convenient space, when it should have received consideration before the larger buildings were all located.

In caring for the various classes of live stock, the question of labor is always an important item, and the class that requires the closest attention to petty details, as a rule, requires the greatest amount of labor. As poultry keeping is wholly a business of details, the economy of labor in performing the necessary work is of great importance. Buildings not conveniently located and arranged become expensive on account of unnecessary labor.

As it is necessary to visit poultry houses several times each day in the year, convenience is of more importance than in case of almost any other farm building. The operations must be performed frequently, so that any little inconvenience in the arrangements of the buildings will cause not only extra expense in the care, but in many cases a greater or less neglect of operations that ought to be performed carefully each day.

Poultry houses are likely to be more or less infested with rats and mice, unless some means are provided to exclude them, and this should be taken into account in selecting a location. It is generally best to locate the poultry house at some distance from other farm buildings, especially if grain is kept in the latter. Convenience of access and freedom from vermin are two desirable points to be secured, and they depend largely upon the location. Everything considered, it is safest to have the house quite isolated.

A dry, porous soil is always to be preferred as a site for buildings and yards. Cleanliness and freedom from moisture must be secured if the greatest success is to be attained. Without doubt, filth and moisture are the causes, either directly or indirectly, of the majority of poultry diseases, and form the stumbling block which brings discouragement and failure to many amateurs. It must not be inferred that poultry can not be successfully reared and profitably kept on heavy soils, for abundant proof to the contrary is readily furnished by successful poultrymen who have to contend with this kind of land. The necessity for cleanliness, however, is not disputed by those who have had extended experience in caring for fowls, particularly the less hardy breeds. That an open, porous soil can be kept comparatively clean with much less labor than a clay soil will be evident to those who are at all acquainted with the habits of domesticated fowls. When the fowls are confined in buildings and yards, that part of the yard nearest the buildings will become more or less filthy from the droppings and continual tramping to which it is subjected. A heavy or clayey soil not only retains all of the manure on the surface, but by retarding percolation at times of frequent showers aids materially in giving to the whole surface a complete coating of filth. If a knoll or ridge can be selected where natural drainage is perfect, the ideal condition will be nearly approached. Where natural favorable conditions as to drainage do not exist, thorough underdrainage will go a long way toward making the necessary amends to insure success.

CONSTRUCTION OF HOUSES.

The material to be used in the construction and the manner of building will necessarily be governed largely by the climatic conditions.

In general, it may be said that the house should provide warm, dry, well-lighted, and well-ventilated quarters for the fowls.

In order to meet these requirements it will be necessary to provide a good roof with side walls more or less impervious to moisture and cold, suitable arrangements for lighting and ventilating, and some means for excluding the moisture from beneath. Where permanent buildings are to be erected, some provision should be made to exclude rats and mice, and for this reason, if for no other, the structure should be placed on cement walls with foundation below the frost line. Cheap, efficient walls may be made of small field stone in the following manner: Dig trenches for the walls below the frost line; drive two rows of stakes in the trenches, one row at each side of the trench, and board inside of the stakes. The boards simply hold the stones and cement in place until the cement hardens. Rough and uneven boards will answer every purpose except for the top ones, which should have the upper edge straight and be placed level to determine the top of the wall. Place two or three layers of stone in the bottom of the trench, put on cement mixed rather thin, and pound down; repeat this operation until the desired height is obtained. The top of the wall can be smoothed off with a trowel or ditching spade and left until the cement becomes hard, when it will be ready for the building.

The boards at the sides may be removed, if desirable, at any time after the cement becomes hard.

For the colder latitudes, a house with hollow or double side walls is to be preferred on many accounts, although a solid wall may prove quite satisfactory, particularly if the building is in the hands of a skilled poultryman. Imperfect buildings and appliances, when under the management of skilled and experienced men, are not the hindrances that they would be to the amateur. Buildings with hollow side walls are warmer in winter and cooler in summer, with less frost in severe weather, and less resulting moisture when the temperature moderates sufficiently to melt the frost from the walls and roof of the house.

A cheap, efficient house for latitudes south of New York may be made of two thicknesses of rough inch lumber for the side and end walls. This siding should be put on vertically, with a good quality of tarred building paper between. In constructing a building of this kind, it is usually best to nail on the inner layer of boards first; then put on the outside of this layer the building paper in such a manner that the whole surface is covered. Where the edges of the paper meet, a liberal lap should be given, the object being to prevent as far as possible drafts of air in severe weather. Nail the second thickness of boards on the building paper so as to break joints in the two boardings. In selecting lumber for siding, it is best to choose boards of a uniform width to facilitate the breaking of joints.

In constructing a roof for a house in the colder latitudes one of two courses must be pursued, either to ceil the inside with some material to exclude drafts or to place the roof boards close together and cover thoroughly with tarred paper before shingling. The ordinary shingle roof is too open for windy weather when the mercury is at or below the zero mark. The fowls will endure severe weather without suffering from frosted combs or wattles if there are no drafts of air. Hens will lay well during the winter months if the houses are warm enough so that the single-comb varieties do not suffer from frost bite. Whenever the combs or wattles are frozen, the loss in decreased egg production can not be other than serious.

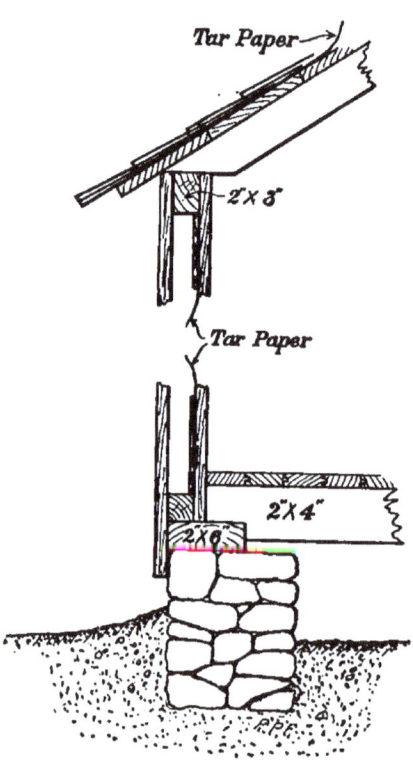

FIG. 1.—Method of building a poultry house with hollow side wall.

Figure 1 represents a cheap and efficient method of building a poultry house with a hollow side wall. The sill may be a 2 by 6 or 2 by 8 scantling, laid flat on the wall or foundation; a 2 by 2 strip is nailed at the outer edge to give the size of the space between the boards which constitute the side walls. A 2 by 3 scantling set edgewise forms the plate, and to this the boards of the side walls are nailed. These boards may be of rough lumber if economy in building is desired. If so, the inner boarding should be nailed on first and covered with tarred building paper on the side that will come within the hollow wall when the building is completed. This building paper is to be held in place with laths or strips of thin boards. If only small nails or tacks are used, the paper will tear around the nail heads when damp and will not stay in place.

The cracks between the boards of the outside boarding may be covered with inexpensive battens if they are nailed at frequent intervals with small nails. Ordinary building lath will answer this purpose admirably, and will last many years, although they are not so durable as heavier and more expensive strips. The tarred paper on the inside boarding and the battens on the outside make two walls, each impervious to wind, with an air space between them.

In preparing plans for a building, one of the first questions to be decided upon is the size and form of the house. If the buildings are made with the corners right angles, there is no form so economical as a square building. This form will inclose more square feet of floor space for a given amount of lumber than any other, but for some reasons a square building is not so well adapted for fowls as one that is much longer than wide. It is essential to have the different pens or divisions in the house so arranged that each one will receive as much sunlight as possible, and to secure this, some sacrifice in economy of building must be made.

The writer prefers a building one story high, and not less than 10 nor more than 14 feet wide, and as long as circumstances require. In most cases a building from 30 to 60 feet long meets all requirements. If this does not give room enough, it is better to construct other buildings than to extend one building for more than 60 feet. It must be remembered that each pen in the building should have a separate yard or run, and that a pen should not be made to accommodate more than 50 fowls, or, better, 30 to 40.

The building should extend nearly east and west in order that as much sunshine as possible may be admitted through windows on the south side. The windows should not be large nor more than one to every 8 or 10 feet in length for a house 12 feet wide, and about 17 inches from the floor, or at such height that as much sunshine as possible will be thrown on the floor. The size and form of the windows will determine quite largely their location. In all poultry houses in cold latitudes the windows should be placed in such a position that they will give the most sunshine on the floor during the severe winter months. One of the common mistakes is in putting in too many windows. While a building that admits plenty of sunlight in the winter time is desirable, a cold one is equally undesirable, and windows are a source of radiation at night unless shutters or curtains are provided. Sliding windows are preferred on many accounts. They can be partially opened for ventilation on warm days. The base or rail on which the window slides should be made of several pieces fastened an inch or so apart, through which openings the dirt which is sure to accumulate in poultry houses may drop and insure free movement of the window.

VENTILATION.

Some means of ventilating the building should be provided. A ventilator that can be opened and closed at the will of the attendant will give good results if given proper attention, and without attention no ventilator will give the best results. All ventilators that are in continuous operation either give too much ventilation during cold and windy weather or not enough during still, warm days. As a rule, they give too much ventilation at night and too little during the warm parts of the day. Ventilators are not needed in severe cold weather, but during the first warm days of early spring, and whenever the temperature rises above freezing during the winter months some ventilation should be provided. Houses with single walls will become quite frosty on the inside during severe weather, which will cause considerable dampness whenever the temperature rises sufficiently to thaw out all the frost of the side walls and roof. At this time a ventilator is most needed. A ventilator in the highest part of the roof that can be closed tightly by means of cords or chains answers the purpose admirably and may be constructed with little expense. The ease and convenience of operation are important points, and should not be neglected when the building is being constructed. It is a simple matter for the attendant to open or close a ventilator as he passes through the house if the appliances for operating it are within easy reach. Figure 2 represents an efficient and easily operated ventilator.

PERCHES.

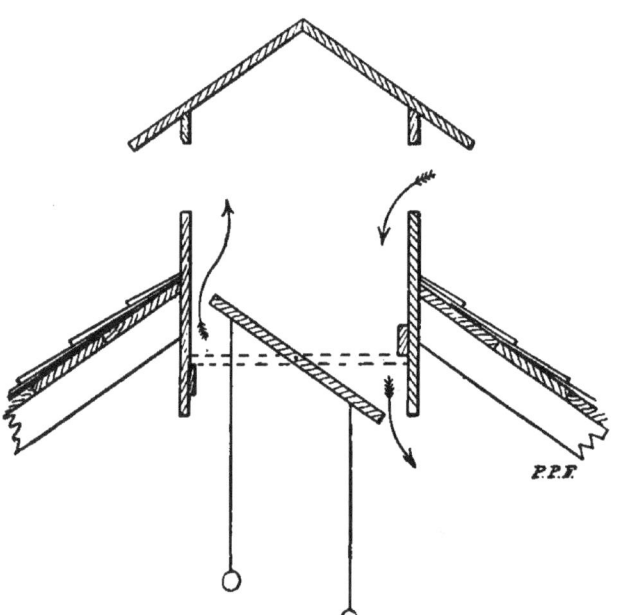

Fig. 2.—Ventilator for a poultry house.

Perches should be not more than 2½ feet from the floor, and should all be of the same height. Many fowls prefer to perch as far above the ground as possible, in order, without doubt, to be more secure from their natural enemies; but when fowls are protected artificially from skunks, minks, foxes, etc., a low perch is just as safe and a great deal better for the heavy-bodied fowls. It must be borne in mind that the distance given at which perches should be placed from the floor applies to all breeds of fowls. It is true that some of the Mediterranean fowls would not in any way be injured in flying to and from the perches, but some of the heavy breeds would find it almost impossible to reach high perches and would sustain positive injuries in alighting on the floor from any considerable elevation. Convenient walks or ladders can be constructed which will enable the large fowls to approach the perches without great effort, but there are always times when even the most clumsy fowls will attempt to fly from the perch to the floor and come down with a heavy thud, which is often injurious. And furthermore, ladders or stairs for the easy ascent of fowls are more or less of a nuisance in the poultry house. The ideal interior arrangement of the house is to have everything that is needed in as simple a form as possible and not to complicate the arrangement by any unnecessary apparatus. The fewer and simpler the interior arrangements the easier the house can be kept clean, and the greater the floor space available for the fowls.

There is no reason why all perches should not be placed near the floor. Movable perches are to be preferred. A 2 by 3 scantling set edgewise, with the upper corners rounded, answers every purpose and makes a satisfactory perch. The perches should be firm and not tip or rock. The form of the scantling makes it easy to secure them firmly and still have them removable.

Underneath the perches should always be placed a smooth platform to catch the droppings. This is necessary for two reasons: the droppings are valuable for fertilizing purposes and ought not to be mixed with the litter on the floor; then, too, if the droppings are kept separate and in a convenient place to remove, it is much easier to keep the house clean than when they are allowed to become more or less scattered by the tramping and scratching of fowls. The distance of the platform from the perch will be governed somewhat by the means employed for removing the droppings. If a broad iron shovel with a tolerably straight handle is used, the space between the platform and perches need not be more than 6 inches. The droppings should be removed every day.

NESTS.

In constructing nest boxes, three points should be kept constantly in mind: (1) The box should be of such a nature that it can be readily cleaned and thoroughly disinfected; if it is removable so that it can be taken out of doors, so much the better; (2) it should be placed in the dark, or where there is only just sufficient light for the fowl to distinguish the nest and nest egg; (3) there should be plenty of room on two or three sides of the nest. It is a well-known fact that some hens in seeking a nest will always drive off other hens, no matter how many vacant nests may be available. If the nest is so arranged that it can be approached only from one side, when one hen is driving another from the nest there is likely to be more or less of a combat, the result of which is often a broken egg. This, perhaps, more than any other one thing, leads to the vice of egg eating. To the writer's knowledge, the habit of egg eating is not contracted where the nests are arranged in the dark and open on two or three sides. Nests for Leghorns or Hamburgs may be made of 6-inch fence boards nailed together so as to form boxes 8 by 10 inches and 6 inches deep. Where perches are arranged with the platform underneath to catch the droppings, as previously described, the nests may be placed on the floor underneath this platform, the opening in front closed with a door which either lets down from the top or lifts from the bottom. Where nests are placed side by side it is necessary to have the partitions between them of sufficient height so that it will be impossible for a hen to draw eggs from one nest to another. Whenever the nest boxes are filled so full with nest material that a hen can draw an egg from one nest to another some of the eggs are likely to be broken.

DRINKING FOUNTAINS.

One of the difficult problems for the poultryman to solve is how to easily provide pure, fresh water for his fowls. Many patent fountains which are on the market are automatic and keep before the fowls a certain quantity of water. Under certain conditions these fountains serve an admirable purpose. Under more adverse conditions many of these patent contrivances fail to give satisfaction for the simple reason that it is impossible to keep them clean. If fowls were fed only whole grain and the weather was always cool, it would be a comparatively easy matter to provide satisfactory automatic drinking fountains, but as soft food forms a considerable portion of the diet for laying hens and fattening fowls, these fountains are necessarily more or less fouled and in warm weather soon become unfit for use as drinking fountains on account of the tainted water and disagreeable odor.

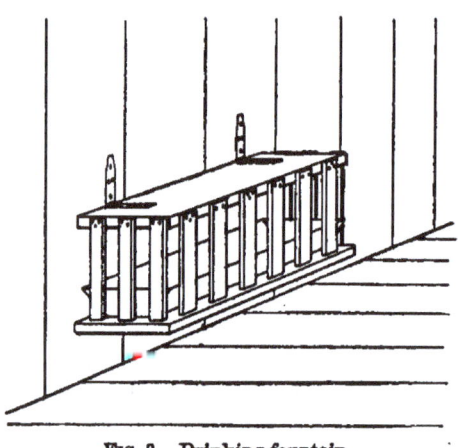

FIG. 3.—Drinking fountain.

A simple, wholesome arrangement may be made as follows: Place an ordinary milk pan on a block or shallow box, the top of which shall be 4 or 5 inches from the floor. The water or milk to be drunk by the fowl is to be placed in this pan. Over the pan is placed a board cover supported on pieces of lath about 8 inches long, nailed to the cover so that they are about 2 inches apart, the lower ends resting upon the box which forms the support of the pan. In order to drink from the pan it will be necessary for the fowls to insert their heads between these strips of lath. The cover over the pan and the strips of lath at the sides prevent the fowl from fouling the water in any manner, except in the act of drinking. Where drinking pans of this kind are used, it is very easy to cleanse and scald them with hot water as occasion demands. This arrangement can be carried a little further by placing a pan, or, what would be still better, a long narrow dish, something like a tin bread tray, on a low shelf a few inches from the floor, and hinging the cover to one side of the poultry house so that it can be tipped up in front for the removal of the dish or for filling it with water. (See fig. 3.) Whatever device is used, it must be easily cleaned and of free access to the fowls at all times.

DUST BOXES.

It is necessary to provide dust boxes for the fowls during the winter months if they are to be kept free from lice. If the soil in the yards is naturally dry and porous, abundant opportunities will be had for dust baths during the warm summer months, but during the late fall, winter, and early spring some artificial provision must be made. A comparatively small box will answer the purpose if the attendant is willing to give a little attention to it each day. These boxes should be placed so that they will receive some sunshine on each bright day, and be kept well filled with loose fine earth. Road dust procured during the hot, dry months of July and August from much-traveled roads has no superior for this purpose. Probably there is no way in which the poultry man can better combat the body louse than by providing dust boxes for his fowls.

YARDS OR PARKS.

Where fowls are kept in confinement it will be found best to provide outdoor runs or yards for them during the summer months. Give them free access to these yards whenever the weather will permit. The most economical form, everything considered, for a poultry yard is one much longer than wide. Two rods wide and 8 rods long is sufficient for 50 fowls. Whenever a poultry plant of considerable size is to be established it will be found most economical to arrange the yards side by side, with one end at the poultry house. The fences which inclose these yards may be made of poultry netting or pickets, and should be at least 7 feet high. In either case it is best to have a board at the bottom, for sometimes it will be desirable to give quite young chickens the run of these yards. If the poultry yards are constructed as described, there is sufficient room for a row of fruit trees down the center of the yard, and still leave ample room for horse cultivation on either side, either with one or with two horses.

These yards are to be kept thoroughly cultivated. If thought best, grain may be sown before cultivation to furnish part of the green food for the fowls. Of all fruit trees, probably there are none that are more suitable for the poultry yard than the plum. The droppings of the fowls will manure the trees, and the fowls as insect destroyers perform a great office in protecting plums from the curculio. After the trees are once well established, a crop of plums should be secured nearly every year. These, too, will require no extra cultivation. The plum trees perform a valuable service in providing shade for the fowls. Where trees are not available, sunflowers may be used for this purpose with a considerable degree of satisfaction. However, some protection must be given the plants until they are well established, and even then many plants will be destroyed unless the fowls have an abundance of green food all the time.

Hamburgs and Leghorns, if they are frequently moved from one pen to another, will sometimes give the owner considerable trouble in flying over fences, even though they are 7 feet high. If it is possible to place the fowls when they are quite young in the yard where they are to remain, much less trouble will be experienced. It has often been noticed that hens would remain peacefully in the yard where they had been reared, but if moved to another yard would give the owner more or less trouble by flying over the inclosure.

SELECTION OF BREEDS AND BREEDING.

A mistake is oftentimes made in selecting fowls of a breed that is not suited for the purposes for which they are to be kept. If egg production is the all-important point, it is a most serious mistake to select a breed of fowls that is not noted for this product. If, on the other hand, meat is the chief object, an expensive mistake will be made if any but the heavy-bodied fowls are chosen. The small, active, nervous, egg-producing breeds can not compete with the larger phlegmatic Asiatics for meat production. Then, too, if fowls are kept for both eggs and meat production, some breed of the middle class should be chosen. These, while they do not attain the great size of the Asiatics, are sufficiently large to be reared profitably to supply the table with meat, and at the same time have the tendency for egg production developed sufficiently to produce a goodly number of eggs during the year. The Wyandottes and Plymouth Rocks are good illustrations of this class of fowls. While individuals of these breeds have made excellent records in egg production, the records of large numbers do not compare favorably with the egg production of the Mediterranean fowls. All of the so-called Mediterranean fowls have a great tendency toward egg production and require only the proper food and care to produce eggs in abundance.

A serious mistake is also made in selecting fowls for breeding purposes and in selecting eggs for hatching. On many farms the custom is to select eggs for hatching during the spring months, when nearly all of the fowls are laying. No matter how poor a layer a hen may be, the chances are that most of the eggs will be produced during the spring and early summer months. A hen that has laid many eggs during the winter months is quite likely to produce fewer eggs during the spring and early summer months than one that commenced to lay on the approach of warm weather. Springtime is nature's season for egg production. All fowls that produce any considerable number of eggs during the year are likely to be laying at this time. It is therefore plain that whenever eggs are selected in the springtime from a flock of mixed hens, composed of some good layers and some poor ones, a larger per cent of eggs will be obtained from the poor layers than at almost any other season of the year. A serious mistake is therefore made in breeding largely from the unprofitable fowls. Whenever it is possible, fowls that are known for the great number of eggs they have produced during the year should be selected for the breeding pen. While it will be almost impossible, and certainly impracticable, in the majority of cases to keep individual records of egg production, yet a selection may be made that will enable the breeder to improve his flock greatly.

The two things necessary to produce large quantities of eggs with the Mediterranean fowls are: (1) Proper food and care, and (2) a strong constitution, which will enable the fowls to digest and assimilate a large amount of food; in other words, fowls so strong physically that they will stand forcing for egg production. In this relation, we may look at the fowl as a machine. If that machine is so strong that it can be run at its full capacity all the time, much greater profit will be derived than if it can be run at its full capacity only a part of the time.

There is, perhaps, no time in the history of the fowl that will indicate its vigor so well as the molting period. Fowls that molt in a very short time and hardly stop laying during this period, as a rule, have strong, vigorous constitutions, and if properly fed give a large yearly record. On the other hand, those that are a long time molting have not the vigor and strength to digest and assimilate food enough to produce the requisite number of eggs. If it is necessary to select fowls at sometime during the year other than the molting period, some indication of their egg-producing power is shown in their general conformation. In selecting a hen for egg production, her form will give some indication of value. A long, deep-bodied fowl is to be chosen rather than one with a short body, whose underline is not unlike a half circle. A strong, hearty, vigorous fowl usually has a long body, a deep chest, with a long and quite straight underline. Other things being equal, the larger bodied fowls of the egg breeds are to be preferred. It is a rule that fowls bred for egg production are larger bodied than those bred for fancy points. Whenever vigor and constitution form an important part in the selection of fowls for breeding, the size of the fowls is invariably increased.

FEEDING.

In feeding for egg production, a valuable lesson may be learned from nature. It will be observed that our domestic fowls that receive the least care and attention, or, in other words, whose conditions approach more nearly the natural conditions, lay most of their eggs in the springtime. It is our duty, then, as feeders, to note the conditions surrounding these fowls at that time. The weather is warm, they have an abundance of green food, more or less grain, many insects, and plenty of exercise and fresh air. Then, if we are to feed for egg production, we will endeavor to make it springtime all the year round; not only to provide a warm place for our fowls and give them a proper proportion of green food, grain, and meat, but also to provide pure air and plenty of exercise.

Farmers who keep only a small flock of hens, chiefly to provide eggs for the family, frequently make a mistake in feeding too much corn. It has been clearly proven by experiment that corn should not form a very large proportion of the grain ration for laying hens; it is too fattening, especially for hens kept in close confinement. Until the past few years, corn has been considered the universal poultry food of America. This, no doubt, has been largely brought about by its cheapness and wide distribution. The recent low prices of wheat have led farmers to feed more of this grain than formerly, and with a consequent improvement in the poultry ration.

When comfortable quarters are provided for the fowls, the nutritive ratio of the food should be about 1:4; that is, one part of protein or muscle-producing compounds to four parts of carbohydrates or heat and fat-producing compounds. Wheat is to be preferred to corn. Oats make an excellent food, and perhaps come nearer the ideal than most any other single grain, particularly if the hull can be removed.

Buckwheat, like wheat, has too wide a nutritive ratio if fed alone, and produces a white flesh and light-colored yolk if fed in very large quantities. In forcing fowls for egg production, as in forcing animals for large yields of milk, it is found best to make up a ration of many kinds of grain. This invariably gives better results than one or two kinds of grain, although the nutritive ratio of the ration may be about the same. It has been found by experiment that the fowls not only relish their ration more when composed of many kinds of grain, but that a somewhat larger percentage of the whole ration is digested than when it is composed of fewer ingredients. It has been clearly proven by experiment that food consumed by the fowls influences the flavor of the eggs; that in extreme cases not only is the flavor of the food imparted to the eggs, but also the odor. This of itself is sufficient reason for always supplying wholesome food for the fowls and seeing to it that none but wholesome food is consumed.

It is conceded by the majority of poultrymen that ground or soft food should form a part of the daily ration. As the digestive organs contain the least amount of food in the morning, it is desirable to feed the soft food at this time, for the reason that it will be digested and assimilated quicker than whole grain. A mixture of equal parts, by weight, of corn and oats ground, added to an equal weight of wheat bran and fine middlings, makes a good morning food if mixed with milk or water, thoroughly wet without being sloppy. If the mixture is inclined to be sticky the proportion of bran should be increased. A little linseed meal will improve the mixture, particularly for hens during the molting period, or for chickens when they are growing feathers. If prepared meat scrap or animal meal is to be fed it should be mixed with this soft food in proportion of about 1 pound to 25 hens. It will be necessary to feed this food in troughs to avoid soiling before it is consumed.

The grain ration should consist largely of whole wheat, some oats, and perhaps a little cracked corn. This should be scattered in the litter which should always cover the floor of the poultry house. It is necessary to have the floor of the poultry house covered with a litter of some kind to insure cleanliness. Straw, chaff, buckwheat hulls, cut cornstalks, all make excellent litters. The object of scattering the grain in this litter is to give the fowls exercise. All breeds of fowls that are noted for egg production are active, nervous, and like to be continually at work. How to keep them busy is a problem not easily solved. Feeding the grain as described will go a long way toward providing exercise. If the fowls are fed three times a day they should not be fed all they will eat at noon. Make them find every kernel. At night, just before going on the perches, they should have all they will eat up clean. At no time should mature fowls be fed more than they can eat. Keep them always active, always on the lookout for another kernel of grain.

GREEN FOOD.

While perhaps not strictly necessary for their existence, some kind of green food is necessary for the greatest production of eggs. Where fowls are kept in pens and yards throughout the year, it is always best to supply some green food. The question how to supply the best food most cheaply is one that each individual must solve largely for himself. In a general way, however, it may be said that during the winter and early spring months, mangel-wurzels, if properly kept, may be fed to good advantage. The fowls relish them, and they are easily prepared. As it is not difficult to grow from 10 to 20 tons of these roots per acre, their cost is not excessive. In feeding these beets to flocks of hens, a very good practice is simply to split the root lengthwise with a large knife. The fowls will then be able to pick out all of the crisp, fresh food from the exposed cut surface. These large pieces have the advantage over smaller pieces in this respect: The smaller pieces when fed from troughs or dishes, will be thrown into the litter and soiled more or less before being consumed by the fowls, and, in fact, many pieces will become so dirty that they will not, nor should they, be eaten. Large pieces can not be thrown about, and remain clean and fresh until wholly consumed.

Clover, during the early spring, is perhaps one of the cheapest and best foods. It is readily eaten when cut fine in a fodder cutter, and furnishes a considerable amount of nitrogen. If clover is frequently mowed, fresh food of this kind may be obtained nearly all summer, particularly if the season be a wet one. Should the supply of clover be limited or the season unusually dry, green food may be cheaply and easily grown in the form of Dwarf Essex rape. This should be sown in drills and given the same cultivation as corn or potatoes. When the rape is from 8 inches to a foot in height, it may be cut and fed. It furnishes a fresh, crisp food that is readily eaten. If cut a few inches from the ground, a second and sometimes a third crop will be produced from one seeding. Alfalfa will also furnish an abundance of green food. It must, however, be cut frequently, each cutting being made before the stalks become hard or woody.

A good quality of clover hay cut fine and steamed makes an excellent food for laying hens if mixed with the soft food.

Cabbages can be grown cheaply in many localities and make excellent green food so long as they can be kept fresh and crisp. Kale and beet leaves are equally as good and are readily eaten. Sweet apples are also suitable, and, in fact, almost any crisp, fresh, green food can be fed with profit. The green food, in many instances, may be cut fine and fed with the soft food, but, as a rule, it is better to feed separately during the middle of the day, in such quantities that the fowls have about all they can eat at one time.

GRIT.

It is necessary that fowls have access to some kind of grit if grain food is fed in any considerable quantities. During the summer months, when they have free access to the yards or runs, it will not be necessary to provide grit, providing the soil is at all gravelly. If, on the other hand, the soil is fine sand or clay, it will be necessary not only to provide grit during the winter months, but throughout the whole year.

Small pieces of crushed stone, flint, or crockery ware will answer the purpose admirably. There are many poultry supply houses which keep constantly on hand crushed granite in various sizes suitable for nearly all kinds of domesticated fowls.

Crushed oyster shells, to a large extent, will supply the necessary material for grinding their food and at the same time furnish lime for the egg shells. Chemical analysis and experiments, together with the reports from many practical poultrymen, show conclusively that the ordinary grain and the green food supplied to laying hens do not contain enough lime for the formation of the egg shells. It will require several times as much lime as is ordinarily fed if good, strong egg shells are to be produced. Crushed oyster shells will supply this necessary lime if kept continually before the fowls, trusting to them to eat the amount needed to supply lime rather than mixing the shells with food. The judgment of the fowl can be relied upon in this respect.

MEAT FOOD.

Where fowls are kept in confinement it will be necessary to supply some meat food. Finely cut fresh bone from the meat markets is one of the best if not the best kind of meat food for laying hens and young chickens. Unfortunately, it is not practicable for many poultrymen to depend wholly on this product, for the reason that it is often inconvenient or impossible to obtain, and when once secured it can not be kept in warm weather without becoming tainted. Tainted bones should be rejected as unfit for food. Skim milk may be substituted wholly or in part for meat food without a decrease in egg production provided the the proper grain ration is given.

FEEDING SMALL CHICKENS.

Chickens do not require food for the first twelve to thirty-six hours after hatching. One of the best foods that can be fed the first few days is stale bread soaked in milk. This should be crumbled fine and placed where the chickens have free access to it, and where they can not step on it. One of the difficult problems for the amateur poultryman is to devise some means for feeding little chickens so that they can consume all of the food without soiling it. If placed on the floor of the brooder or the brooder run, the larger part of the food will be trampled upon and will soon become unfit to eat.

A simple and efficient feeding trough may be made by tacking a piece of tin about $3\frac{1}{2}$ inches wide along the edge of a half-inch board so that the tin projects about an inch and a half on either side of the board, bending the tin so as to form a shallow trough, and fastening the board to blocks which raise it from 1 to 2 inches from the floor. (See fig. 4.) The trough may be from 1 to 3 feet long. It

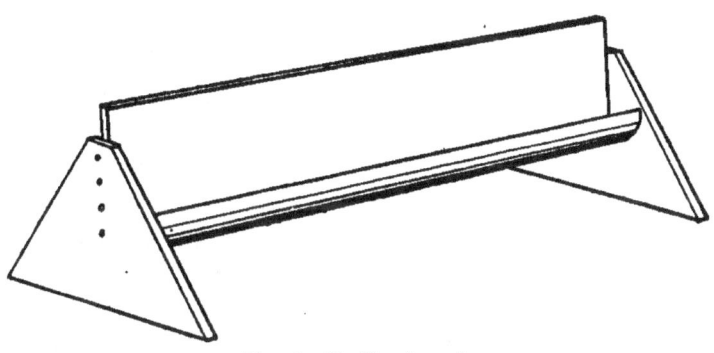

FIG. 4.—Feeding trough.

is within easy reach of the chickens and so narrow that they can not stand upon the edges. Food placed in such feeding troughs can be kept clean until wholly consumed.

Granulated oats (with the hulls removed) make an excellent food for young chickens. There is, perhaps, no better grain food for young chickens than oats prepared in this manner. It may be fed to good advantage after the second or third day in connection with the bread sopped in milk. A good practice is to keep it before them all the time.

The chickens should have free access to some kind of grit after the first day. Coarse sand makes an excellent grit for very young chickens. As they get a little older some coarser material must be provided.

Milk is an excellent food for these young fowls, but requires skill in feeding.

One of the great difficulties in rearing fowls is to carry young chickens through the first two weeks without bowel disorders. Too low temperature in the brooder, improper food and injudicious feeding, even if the right kinds of food are given, each plays an important part in producing these disorders. After the first ten days milk may be given more freely, perhaps, than during the earlier stages of the chick's existence. As the chick becomes a little older, more uncooked food may be fed. A mixture of fine middlings, wheat bran, a little corn meal, and a little linseed meal mixed with milk makes a valuable food. Hard-boiled eggs may be fed from the beginning, but, like milk, require more skill than the feeding of bread sopped in milk. On farms where screenings from the various grains become really a by-product, these form a cheap and efficient food for the little chickens. Wheat screenings, especially, form one of the best foods, particularly if they contain a considerable portion of good kernels that have been cracked in thrashing. Then, too, the screenings contain a number of weed seeds that have some feeding value and are relished by the fowls. They not only provide sustenance, but give variety, and this, in a measure, improves the general health.

Drinking fountains require close attention. Small chickens drink frequently and oftentimes with their beaks loaded with food, which is left, to a greater or less extent, in the water supply. As it is necessary to keep these fountains in a tolerably warm atmosphere, they soon become tainted and emit a disagreeable odor. This condition must not be allowed to exist, for all the food and drink consumed by fowls should be wholesome. It has often been said that "cleanliness is next to godliness," and certain it is that cleanliness is next to success in poultry keeping. The drinking fountains must be kept clean. If automatic fountains are used great care must be exercised in keeping them clean and free from bad odors. Nothing less than frequent scalding with steam or hot water will answer the purpose. A cheap, efficient drinking fountain may be made of a tin can with a small hole in one end near the side of the can, under which is soldered a crescent-shaped piece of tin, forming a lip or a small receptacle for water. If the can is filled with water and then placed on its side, a small quantity of water will run out of the opening and remain in this crescent-shaped lip. As the chicks drink this water a quantity of air will pass into the opening and a little more water will flow out. This kind of fountain will keep before the chickens a small quantity of water at all times accessible. By exercising care and keeping the fountain thoroughly clean, satisfactory results are easily obtained from this arrangement.

BROODERS.

If one resorts to artificial incubation it will be necessary to provide a brooder of some kind. It may be simple and quite inexpensive, or complex and costly. It is not necessary to expend very much money in the construction of an efficient brooder. It is necessary, however, to see that the brooder is capable of doing certain things. Some of these requisites are summed up in the following: It must be warm. The little chickens require a temperature of from 90° to 100° the first few days, and at all times they should find it so warm in the brooder that they are not inclined to huddle together to keep warm. If the brooder is automatic, then the temperature may be kept quite even throughout the whole floor space. If, on the other hand, the brooder is heated from one side or from the top, and is not automatic, it will be best to construct it so that certain parts of the machine will be very warm, in fact, a little warmer than is necessary for the chickens, and some other part somewhat too cool. It does not take them long to learn just where the most comfortable position is. They may be trusted entirely to select the proper temperature if the brooder is of sufficient size so that it is never crowded. A brooder constructed on this plan will require less attention than almost any other. It may undergo a considerable variation in temperature without overheating or chilling the chickens.

The brooder should be easily cleaned and so constructed that all of the floor space can readily be seen. Inconvenient corners are objectionable in brooders, in fact, any corner is objectionable, but if brooders are constructed cheaply, it is almost necessary to make more or less corners. If constructed of wood, circular ones are somewhat more expensive than square or rectangular ones. The floor must not only be kept clean but dry.

Top or side heat is to be preferred to bottom heat, but there must be sufficient bottom heat to keep the floor dry.

As the chickens get a few days old, plenty of exercise must be provided. One objection to many of the brooders in the market is that the chickens are kept too closely confined and not allowed sufficient exercise. It will be a matter of surprise to many to learn how much exercise these little fellows require. With the young chicken, as with the athlete, strength is acquired by exercise, and above all other conditions of growth, strength is the one thing necessary in the young chicken.

INCUBATORS.

The modern improvement in incubators has made the rearing of fowls solely for egg production quite out of the question unless these machines are used. No experienced poultryman at the present time will undertake to rear fowls in large numbers for the production of eggs and depend on the hens that lay the eggs for incubation. The Mediterranean fowls can not be depended upon for natural incubation. Artificial incubation must be resorted to if these fowls are to be reared in considerable numbers.

There are many kinds of excellent incubators on the market. As with many kinds of farm machinery, it is impossible to say that one particular kind is better than all others. Then, too, an incubator that would give very satisfactory results with one individual might prove to be quite inferior in the hands of another person. What is best for one is not necessarily best for another. It is advisable, before investing extensively in any make of incubator, to thoroughly understand the machine. If good results are obtained, then additional machines of the same kind should be purchased. Failures are recorded simply because the individual fails to thoroughly understand the machine he is trying to operate, or, in other words, fails to learn how to operate that particular machine to the best advantage. A successful poultryman must necessarily pay close attention to petty details. Not only is this necessary in caring for little chickens and mature fowls, but also in the care and management of incubators and brooders. The whole business is one of details. While incubators may vary considerably one from another, yet there are certain points to which all should conform. Some of these points are summed up in the following:

(1) They should be well made of well-seasoned lumber. The effort of manufacturers to meet a popular demand for cheap machines has placed on the market incubators that are not only cheaply made, but made of cheap and not thoroughly seasoned material.

(2) The incubator should be easy of operation. All its adjustments should be easily made and so arranged that the more delicate machinery is in plain view of the operator. The machine should be automatic in operation. When supplied with the necessary heat it should control perfectly within certain limits the temperature of the egg chamber. This result is accomplished in various ways. The regulating force, whatever it may be, should be placed within the egg chamber so that the regulator may vary as the temperature in the egg chamber varies, irrespective of the changes of temperature of the room in which the incubator is placed. The regulator must be sensitive. The change of temperature which is necessary for the complete working of the regulator ought not to be more than 1 degree; that is, 1 degree above or below the desired temperature. It is better if the range of temperature can be reduced to one-half of 1 degree, thus making a total variation of 1 degree instead of 2 degrees.

It should not be inferred that a much wider variation than this will not give excellent results under otherwise favorable conditions, but, other things being equal, those machines which are most nearly automatic are to be preferred.

In addition to the foregoing requisites, a convenient appliance for turning the eggs, positive in its action, should accompany each incubator. This may be an extra tray that is to be placed bottom side up over the tray of eggs and held firmly in this position while both trays are turned, thus completely transferring the eggs from one tray to another without jar. The different machines have very different appliances for accomplishing this result. Excellent results are obtained by the use of many machines now on the market when the operator of these various machines is thoroughly interested. Poultrymen have, for a term of years, hatched in incubators over 80 per cent of all eggs put in the machine. It must not be inferred that this is an easy thing to do. A record of this kind is attained only by close observation and good judgment, not only in running the machine, but also in the breeding and care of the fowls to produce fertile eggs.

DISEASE AND LICE.

Disease and lice are the great obstacles to be overcome in poultry raising. The houses may be kept free from lice by a liberal use of kerosene emulsion and by whitewashing. Whitewash serves a double purpose, that of ridding the house of lice and making the interior much lighter. A small window, with the interior of the house whitewashed, will make the building as light as a much larger window without the whitewash. If the poultry houses are kept free from lice, the fowls can usually be depended upon to keep themselves free by a liberal use of the dust bath. If, however, body lice are found, they may be successfully treated by dusting insect powder under the feathers in the evening and allowing the fowls to remain undisturbed on the perches after the treatment.

Gapes in chickens frequently destroy large numbers, and are caused by trematode worms in the windpipe. The number of worms is sometimes so great as to completely choke the fowl. A feather moistened with turpentine or kerosene oil and inserted into the windpipe and turned until the worms are removed is a practice quite largely recommended. Others recommend removing the worms with a fine wire or horsehair, doubled so as to form a loup; this is to be inserted into the windpipe and turned until the worms are detached, and then withdrawn, bringing the worms with it. Another remedy practiced by some poultrymen is to cause the chickens to breathe air in a confined space into which fine, slaked lime is occasionally dusted.

Preventive measures are far more satisfactory than the treatment of infected fowls. The pens and yards should be kept clean and dry and the chickens kept in as thrifty condition as possible by supplying proper food and exercise. While these conditions may not insure absolute freedom from the disease in every instance, yet to moisture and filth can be attributed nearly all cases of gapes, particularly if the yards or pens were previously occupied by infected birds. Yards that have been allowed to become damp, filthy, and infected with the gapeworm may be improved by draining and thorough cultivation. Heavy applications of lime just before cultivating or saturation of the soil with strong salt solution (provided no crop is to be grown) are recommended by experienced poultrymen.

Chicken cholera.—This is an exceedingly fatal contagious disease, which is widely distributed over this country, and causes enormous annual losses, especially in the central and southern sections.

The first symptom of the disease is, in the majority of cases, a yellow coloration of that part of the excrement which is usually white, quickly followed by violent diarrhea and rise of temperature. Other common accompanying symptoms are drooping of the wings, stupor, lessened appetite, and excessive thirst. Since the disease is due to a specific germ, it can only be introduced into a flock by direct importation of this germ, generally by fowls from infected premises. As soon as the symptoms of the disease are observed "the fowls should be separated as much as possible and given restricted quarters, where they may be observed and where disinfectants can be freely used. As soon as the peculiar diarrhea is noticed with any of the fowls, the birds of that lot should be changed to fresh ground and the sick ones killed. The infected excrement should be carefully scraped up and burned, and the inclosure in which it has been thoroughly disinfected with a one-half per cent solution of sulphuric acid or a 1 per cent solution of carbolic acid, which may be applied with an ordinary watering pot. Dead birds should be burned or deeply buried at a distance from the grounds frequented by the fowls.

"The germs of the disease are taken into the system only by the mouth, and for this reason the watering troughs and feeding places must be kept thoroughly free from them by frequent disinfection with one of the solutions mentioned. * * *

"Treatment of sick birds is not to be recommended under any circumstances. The malady runs its course, as a rule, in one, two, or three days, and it can only be checked with great difficulty."[1]

Roup is one of the most dreaded of diseases. It is sometimes spoken of as the winter disease. The symptoms are hoarse breathing, swelled eyes, discharge at the nostrils, and sometimes a fetid breath. Treatment is not generally satisfactory. The affected birds should be removed, the houses cleansed and disinfected. Damp, foul air and cold drafts in the poultry houses should be carefully avoided whenever fowls are subject to roup. A decrease in the proportion of corn and an increase in the proportion of meat food in the daily ration is held by some to be highly beneficial in warding off this disease.

In general, the treatment of the common diseases of fowls is not so satisfactory as preventive measures. Nowhere more than in the poultry business does that old adage apply, "An ounce of prevention is worth a pound of cure."

DRESSING AND SHIPPING.

A considerable proportion of the dressed poultry consigned to commission houses in large cities brings to the producer a much smaller profit than it would had the same poultry been dressed and packed for shipment with greater skill. It is of prime importance that the poultry products be placed on the market in a condition that will make them appear as inviting as possible. Proper feeding for two or three weeks before the fowls are slaughtered will improve their color materially.

---

[1] D. E. Salmon, U. S. Dept. Agr., Rpt. 1880, p. 444.

In most of the American markets fat fowls with a yellow skin bring the highest price. This condition may be secured most cheaply by feeding a grain ration composed largely of corn for two or three weeks before the fowls are slaughtered. Of the more common grain foods there is none that excels corn for this purpose.

The commission men and shippers, who study in detail dressing and packing, state that uniformly fine quality will soon acquire a reputation among buyers. The shipper should always be careful to have the product look as neat as possible. In some of the large cities ordinances prohibit the sale of dressed poultry with food in their crops. In a few instances the sale of live poultry in coops which contain food is also prohibited. In all cases it is best to withhold food from twelve to twenty-four hours before killing, but the fowls should have plenty of water during this time, that they may be able to digest and assimilate food already consumed. All fowls should be killed by cutting through the roof of the mouth and allowing them to bleed to death. In all operations of dressing avoid cutting or bruising the skin or breaking bones. Care is required in the case of the heavy fowls in picking and handling to prevent bruising the skin. In packing fowls use neat, clean, and as light packages as will carry safely. Boxes or barrels holding about two hundred pounds meet these requirements best; boxes are better for turkeys and geese and barrels for chickens. Barrels may be used, however, for dry shipment as well as for hot weather shipment when the fowls are to be packed in ice.

In shipping live poultry the coop should be high enough to allow the fowls to stand upright without bending their legs. When large coops are used there should be partitions, so that if the coops are tipped all of the fowls are not thrown to one side. They should have plenty of room in the coop. If possible put only one kind in a coop or in one division of a coop.

FARMERS' BULLETINS.

These bulletins are sent free of charge to any address upon application to the Secretary of Agriculture, Washington, D. C. Only the following are available for distribution:

No. 16. Leguminous Plants for Green Manuring and for Feeding. Pp. 24.
No. 18. Forage Plants for the South. Pp. 30.
No. 19. Important Insecticides: Directions for Their Preparation and Use. Pp. 20.
No. 21. Barnyard Manure. Pp. 32.
No. 22. Feeding Farm Animals. Pp. 32.
No. 23. Foods: Nutritive Value and Cost. Pp. 32.
No. 24. Hog Cholera and Swine Plague. Pp. 16.
No. 25. Peanuts: Culture and Uses. Pp. 24.
No. 26. Sweet Potatoes: Culture and Uses. Pp. 30.
No. 27. Flax for Seed and Fiber. Pp. 16.
No. 28. Weeds; and How to Kill Them. Pp. 30.
No. 29. Souring of Milk, and Other Changes in Milk Products. Pp. 23.
No. 30. Grape Diseases on the Pacific Coast. Pp. 16.
No. 31. Alfalfa, or Lucern. Pp. 23.
No. 32. Silos and Silage. Pp. 31.
No. 33. Peach Growing for Market. Pp. 24.
No. 34. Meats: Composition and Cooking. Pp. 29.
No. 35. Potato Culture. Pp. 23.
No. 36. Cotton Seed and Its Products. Pp. 16.
No. 37. Kafir Corn: Characteristics, Culture, and Uses. Pp. 12.
No. 38. Spraying for Fruit Diseases. Pp. 12.
No. 39. Onion Culture. Pp. 31.
No. 40. Farm Drainage. Pp. 24.
No. 41. Fowls: Care and Feeding. Pp. 24.
No. 42. Facts About Milk. Pp. 29.
No. 43. Sewage Disposal on the Farm. Pp. 22.
No. 44. Commercial Fertilizers. Pp. 24.
No. 45. Some Insects Injurious to Stored Grain. Pp. 32.
No. 46. Irrigation in Humid Climates. Pp. 27.
No. 47. Insects Affecting the Cotton Plant. Pp. 32.
No. 48. The Manuring of Cotton. Pp. 16.
No. 49. Sheep Feeding. Pp. 24.
No. 50. Sorghum as a Forage Crop. Pp. 24.
No. 51. Standard Varieties of Chickens. Pp. 48.
No. 52. The Sugar Beet. Pp. 48.
No. 53. How to Grow Mushrooms. Pp. 20.
No. 54. Some Common Birds in Their Relation to Agriculture. Pp. 40.
No. 55. The Dairy Herd: Its Formation and Management. Pp. 24.
No. 56. Experiment Station Work—I. Pp. 30.
No. 57. Butter Making on the Farm. Pp. 15.
No. 58. The Soy Bean as a Forage Crop. Pp. 24.
No. 59. Bee Keeping. Pp. 32.
No. 60. Methods of Curing Tobacco. Pp. 16.
No. 61. Asparagus Culture. Pp. 40.
No. 62. Marketing Farm Produce. Pp. 28.
No. 63. Care of Milk on the Farm. Pp. 40.
No. 64. Ducks and Geese. Pp. 48.
No. 65. Experiment Station Work—II. Pp. 32.
No. 66. Meadows and Pastures. Pp. 24.
No. 67. Forestry for Farmers. Pp. 48.
No. 68. The Black Rot of the Cabbage. Pp. 22.
No. 69. Experiment Station Work—III. Pp. 32.
No. 70. The Principal Insect Enemies of the Grape. Pp. 24.
No. 71. Some Essentials of Beef Production. Pp. 24.
No. 72. Cattle Ranges of the Southwest. Pp. 32.
No. 73. Experiment Station Work—IV. Pp. 32.
No. 74. Milk as Food. Pp. 39.
No. 75. The Grain Smuts. Pp. 20.
No. 76. Tomato Growing. Pp. 30.
No. 77. The Liming of Soils. Pp. 19.
No. 78. Experiment Station Work—V. Pp. 32.
No. 79. Experiment Station Work—VI. Pp. 28.
No. 80. The Peach Twig-borer—an Important Enemy of Stone Fruits. Pp. 16.
No. 81. Corn Culture in the South. Pp. 24.
No. 82. The Culture of Tobacco. Pp. 23.
No. 83. Tobacco Soils. Pp. 23.
No. 84. Experiment Station Work—VII. Pp. 32.
No. 85. Fish as Food. Pp. 30.
No. 86. Thirty Poisonous Plants. Pp. 32.
No. 87. Experiment Station Work—VIII. Pp. 32.
No. 88. Alkali Lands. Pp. 23.
No. 89. Cowpeas. Pp. 16.
No. 90. The Manufacture of Sorghum Sirup. (In press.)
No. 91. Potato Diseases and Their Treatment. (In press.)
No. 92. Experiment Station Work—IX. (In press.)
No. 93. Sugar as Food. (In press.)
No. 94. The Vegetable Garden. (In press.)

www.ingramcontent.com/pod-product-compliance
Lightning Source LLC
Chambersburg PA
CBHW062208220526
45470CB00009B/2968